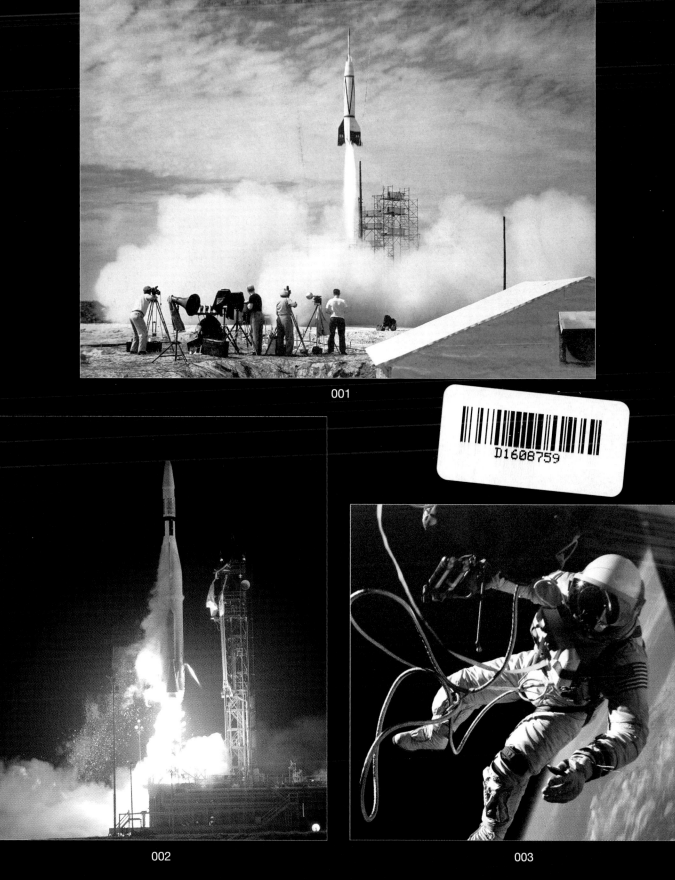

001. First rocket launch from Cape Canaveral, Bumper 2, 1950. **002.** Mariner 1 is lifted into space by Atlas-Agena 5 to orbit Venus, 1962. **003.** Edward H. White II, first American to spacewalk, Gemini IV, 1965.

SPACE EXPLORATION 1

004

005

006

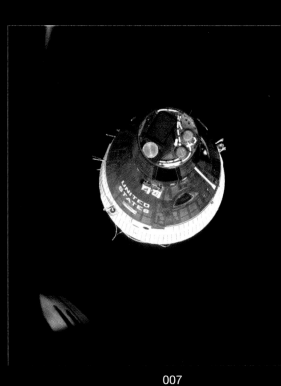
007

004, 005. Edward H. White II, first American to spacewalk, Gemini IV, 1965. **006.** Gemini VI in orbit above Earth as seen from Gemini VII, 1965. **007.** Gemini VII as seen from Gemini VI, 1965.

008

009

010

008. Agena Target Vehicle as seen from Gemini VIII, 1966. **009.** Lunar Module Spider, Apollo 9, 1969. **010.** Apollo 11 launch, the first lunar landing mission, 1969.

011. Neil A. Armstrong by Lunar Module Eagle, Apollo 11, 1969. **012.** Buzz Aldrin during moonwalk, Apollo 11, 1969.
013. Buzz Aldrin next to Solar Wind Composition experiment with Lunar Module Eagle in background, Apollo 11, 1969.

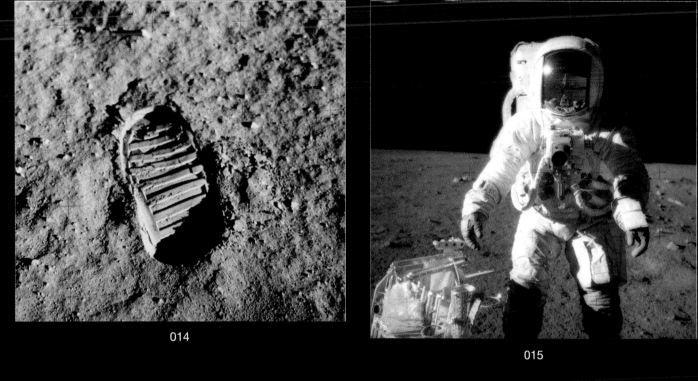

014

015

016

017

014. Buzz Aldrin's boot print, one of the first steps taken on the moon, Apollo 11, 1969. **015.** Alan L. Bean during moonwalk, Charles Conrad Jr. in helmet reflection, Apollo 12, 1969. **016.** Apollo 14 launch from Kennedy Space Center, 1971. **017.** View from Lunar Module of Command-Service Modules, Apollo 15, 1971.

018

019

020

018. Charles M. Duke, Jr. stands by Plum Crater with Lunar Roving Vehicle in background, Apollo 16, 1972. **019.** Eugene A. Cernan driving Lunar Roving Vehicle, Apollo 17, 1972. **020.** Eugene A. Cernan salutes flag on Moon, Lunar Module Challenger behind flag, Lunar Roving Vehicle behind Cernan, Apollo 17, 1972.

021

022

023

021. Harrison Schmitt takes samples of the Moon's crust, Apollo 17, 1972. **022.** Skylab Orbital Workshop in Earth orbit, 1974. **023.** Viking 1 launch from Cape Canaveral, 1975.

024

025

026

024. Viking 1 launch from Cape Canaveral, 1975. **025.** Soviet Soyuz spacecraft in Earth orbit, 1975.
026. Space Shuttle Challenger, 1983.

027

028

029

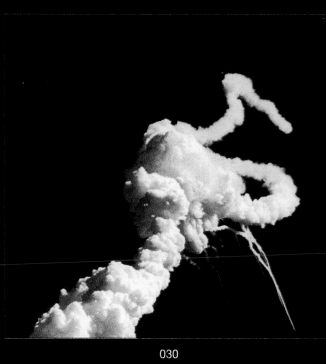

030

027. Bruce McCandless II tests a Mobile Foot Restraint, Space Shuttle Challenger, 1984. **028.** Bruce McCandless II free flying with a nitrogen jet-propelled backpack, Space Shuttle Challenger, 1984. **029.** Jerry L. Ross anchored to foot restraint on Remote Manipulator System, Space Shuttle Atlantis, 1985. **030.** Space Shuttle Challenger disaster, 1986.

031

032

031. Space Shuttle Atlantis liftoff, 1988. **032.** Space Shuttle Discovery launch carrying Hubble Space Telescope with Space Shuttle Columbia on launchpad to the left, 1990.

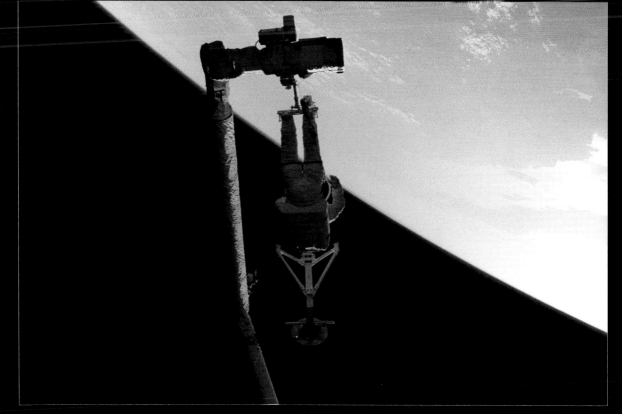

033

034

035

033. Pierre Thuot on the end effector of the Robot Arm holding Intelsat VI capture bar, May 1992. **034.** Pierre Thuot on the end effector of the Robot Arm attempting to capture Intelsat VI, 1992. **035.** G. David Low and Peter J. Wisoff conducting detailed Test Objective procedures, Space Shuttle Endeavour, Earth in background, 1993.

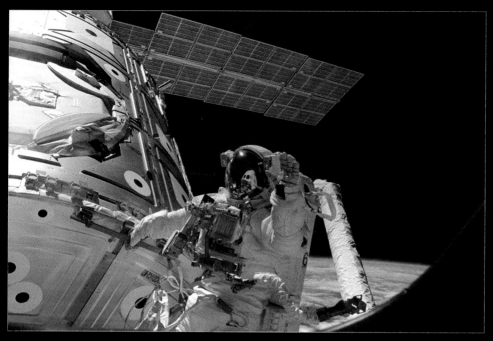

036

037

038

036. James H. Newman waves at camera with Discovery's reflection on helmet, Space Shuttle Discovery, 1993. **037.** James H. Newman evaluates the Portable Foot Restraint with Earth's horizon in background, Space Shuttle Discovery, 1993. **038.** Kathy Thornton removes a damaged panel from the Hubble Space Telescope in its first servicing mission, December 1993.

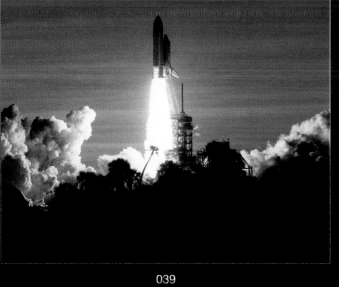

039

040

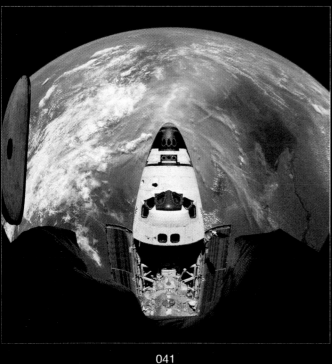

041

042

039. Space Shuttle Discovery launch from Kennedy Space Center, 1994. **040.** Mark C. Lee tests Simplified Aid for EVA Rescue system, Earth in background, 1994. **041.** Fish eye view of Space Shuttle Atlantis as seen from Mir Russian Space Station, 1995. **042.** Space Shuttle Atlantis departs from the Mir Russian Space Station, 1995.

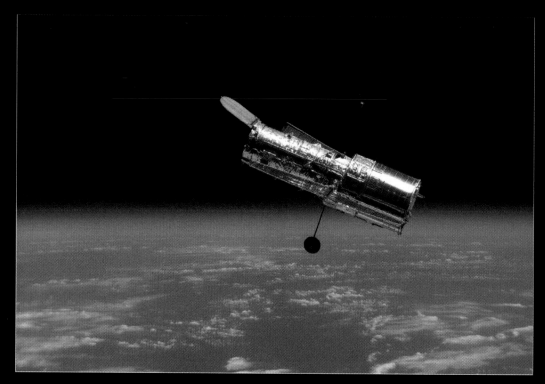

043

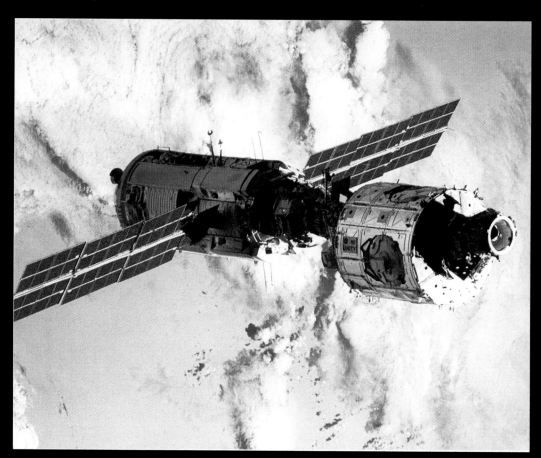

044

043. Hubble Space Telescope hovers over Earth, 1997. **044.** Unity connected to Russian space station component Zarya, 1998.

045

046

047

045. Space Shuttle Discovery approaches Mir Russian Space Station, 1998. **046.** International Space Station moves away from Space Shuttle Discovery, 1998. **047.** Zvezda is linked to Zarya on International Space Station, 2000.

048

049

048. International Space Station viewed from Space Shuttle Discovery, 2001.
049. Space Shuttle Discovery hitches a ride to Kennedy Space Center, 2005.

050

051

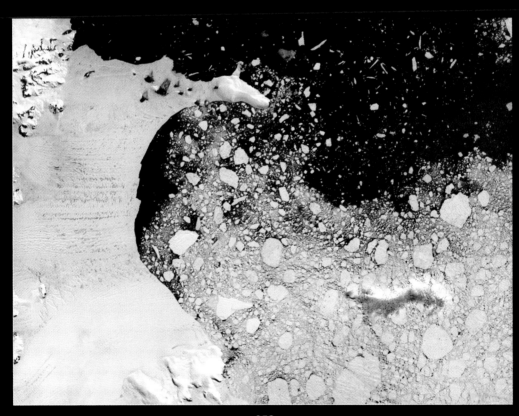

052

050. View of Earth from Apollo 17. **051.** Satellite image of Earth's interrelated systems and climate.
052. Antarctic, view of icebergs splitting from the Larsen Ice Shelf, 2000.

053

054

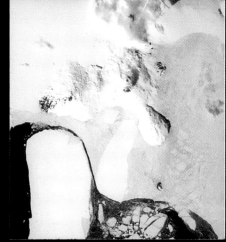

055

056

053. Antarctic, Ross Sea, January 16, 2001.
054. Antarctic, Ross Sea, February 15, 2001. **055.** Antarctic, Ross Sea, December 9, 2001. **056.** Plume of steam or ash from Mount Etna, 2001.

057

058

057. Top of the Earth's atmosphere with the Moon in background. **058.** Hurricane Ivan, 2004.

059. Earth's Meteor Crater, formed approximately 50,000 years ago (1.1 kilometer wide, 200 meters deep).
060. View of lights at night around the world. **061.** Moon sets over Earth's horizon. **062.** Moonrise.

063

064

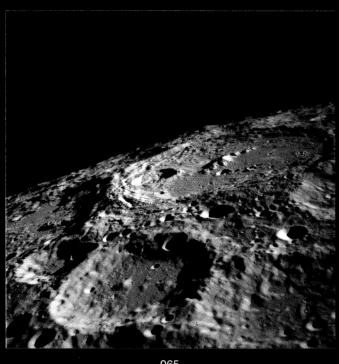

065

063. Moon illuminated by Earthshine, Venus at top. **064.** Moon's North Pole. **065.** Crater 302 on the Moon's surface.

066

067

068

066. Mars. **067.** Panorama of Mars Pathfinder landing site. **068.** Colorized view of Mars from the Mars Exploration Rover Spirit.

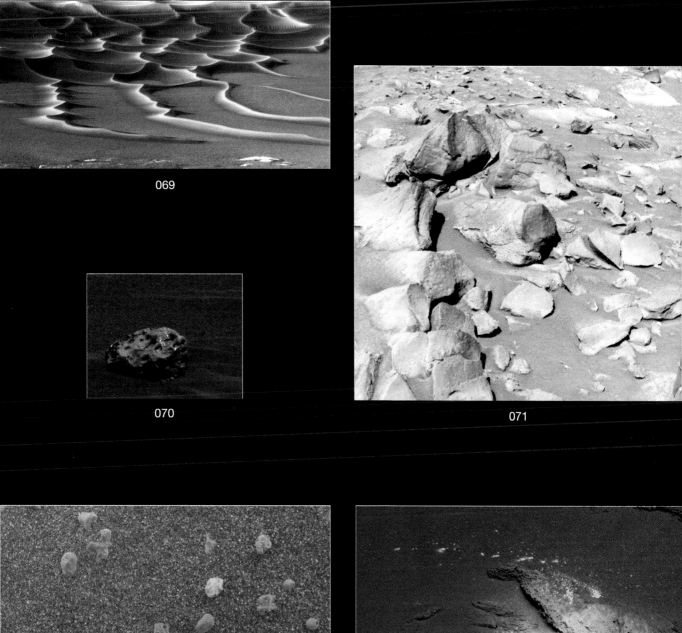

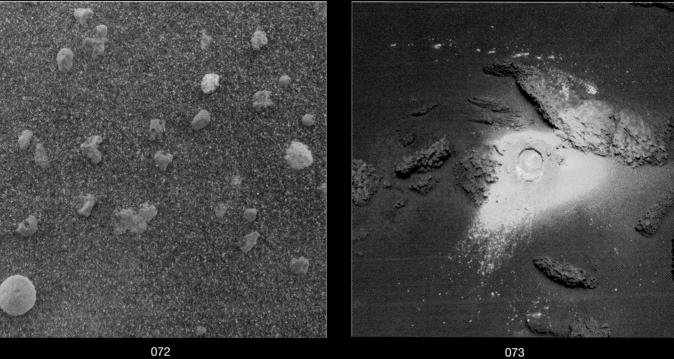

069. Dune field on floor of Endurance Crater on Mars. **070.** Meteorite on Mars, first meteorite identified on another planet. **071.** Basaltic or volcanic rocks on Mars. **072.** Magnified look at soil on Mars. **073.** Sulfur-rich rocks and surface materials on Mars.

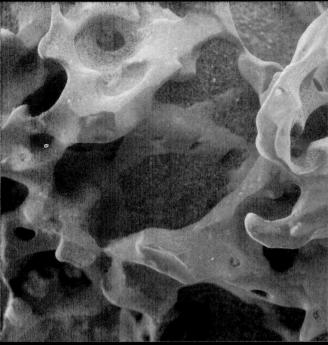

074

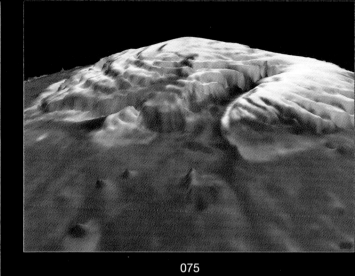

075

076

074. Hardened lava on Mars. **075.** 3-D view of Mars. **076.** Sunset sinking below Gusev crater on Mars.

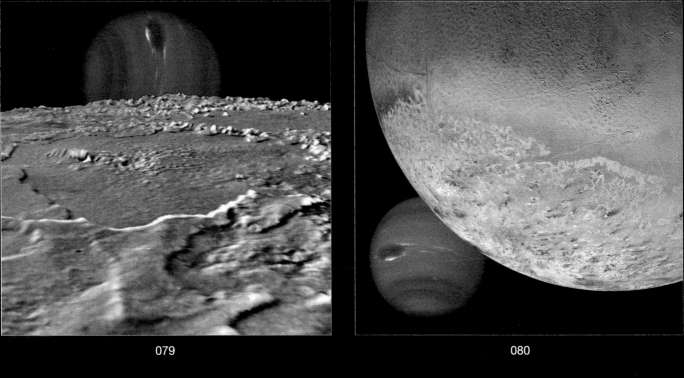

077. Uranus. **078.** Neptune. **079.** Neptune on Triton's horizon. **080.** Triton with Neptune in background.

081

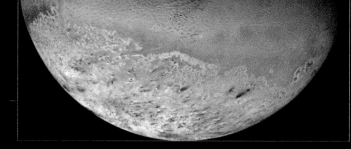

082

083

081, 082. Triton, Neptune's largest satellite. **083.** Saturn.

084

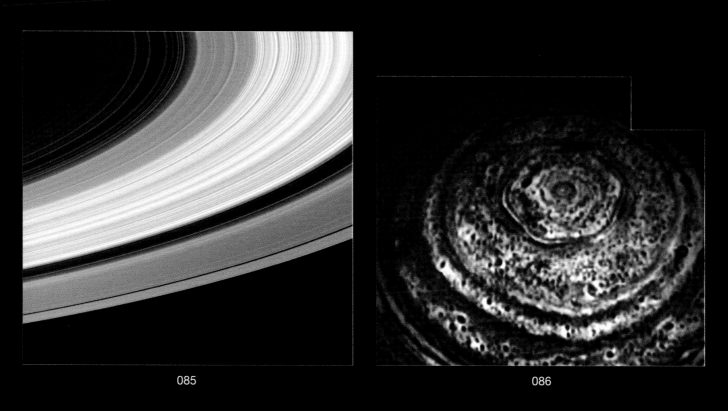

085   086

084, 085. Saturn's rings. **086.** Saturn's North Pole.

087

088

089

087. SATURNIAN SYSTEM: Dione in front, Saturn behind, Tethys and Mimas on right, Enceladus and Rhea to left, Titan at top. **088.** Rhea. **089.** Rhea with Saturn in background.

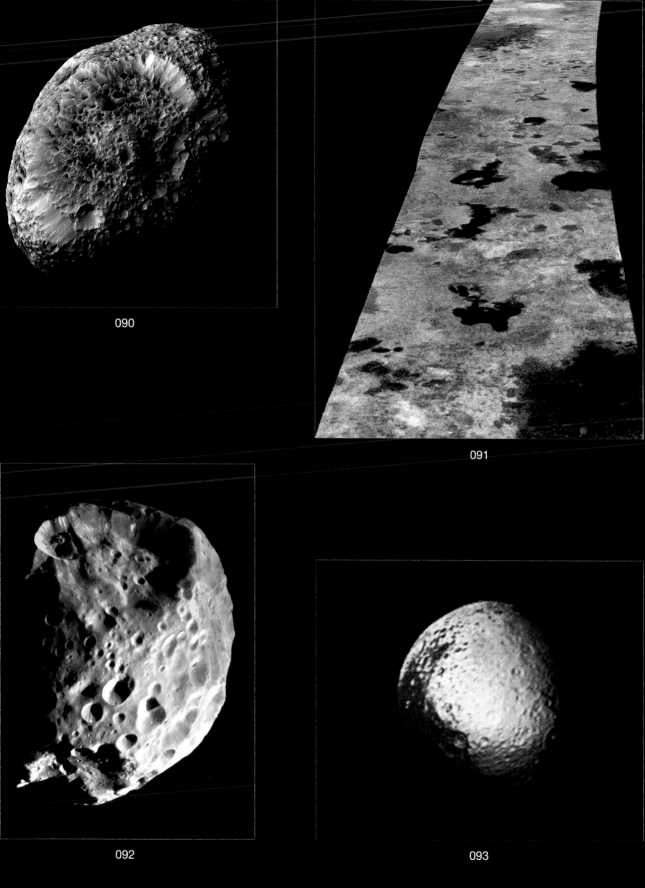

MOONS OF SATURN: **090.** Hyperion. **091.** Titan (colorized image of lakes). **092.** Pheobe. **093.** Iapetus.

094

095 · 096

094–096. Jupiter, photos taken by Cassini spacecraft.

30 PLANETS AND MOONS

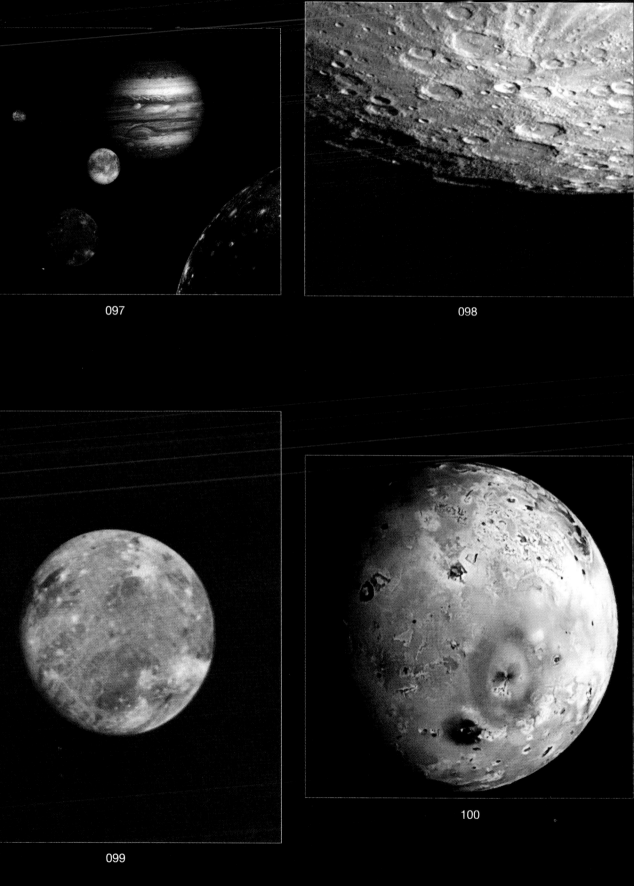

097 JUPITER SYSTEM: Io (upper left), Europa (center), Ganymede and Callisto (lower right). **098.** Mercury's South Pole. **099.** Ganymede, Jupiter's largest moon. **100.** Io, Jupiter's third largest moon, with active volcanoes.

101

102

103

104

101–103. Io, Jupiter's third largest moon, with active volcanoes. **104.** Europa, Jupiter's fourth largest moon.

105

106

105. Europa, Jupiter's fourth largest moon., **106.** LEFT TO RIGHT: Neptune, Uranus, Saturn, Jupiter.

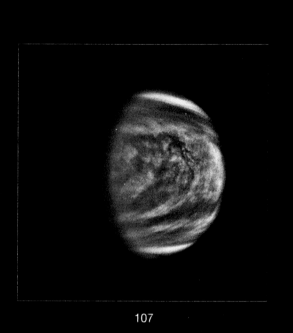

107

108

109

107. Venus cloud patterns. **108.** Top to bottom: Mercury, Venus, Earth with Moon, Mars, Jupiter, Saturn, Uranus, Neptune. **109.** Solar system with four of Jupiter's moons.

110

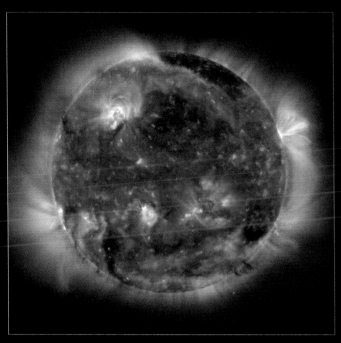

111

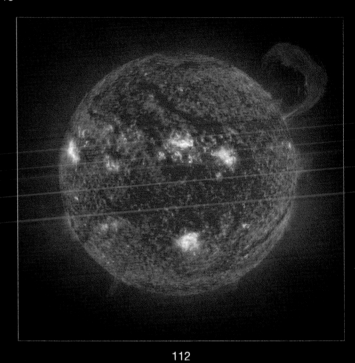

112

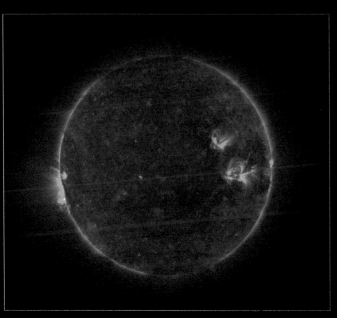

113

110. LEFT TO RIGHT (distance from Sun): Mercury, Venus, Earth, Mars, Jupiter, Saturn, Uranus, Neptune, Pluto. **111.** Color composite of solar features of the Sun. **112.** Sun with a handle-shaped prominence. **113.** Highly charged particles burst from an active region on the Sun's surface.

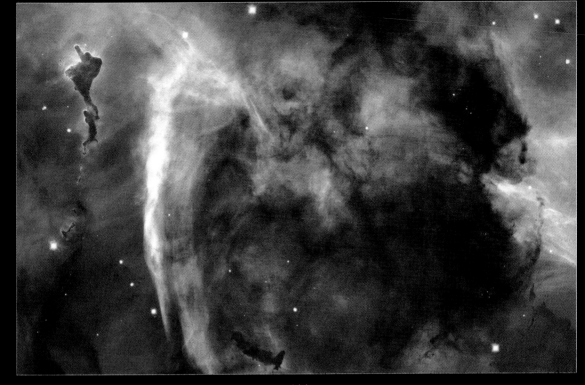

114

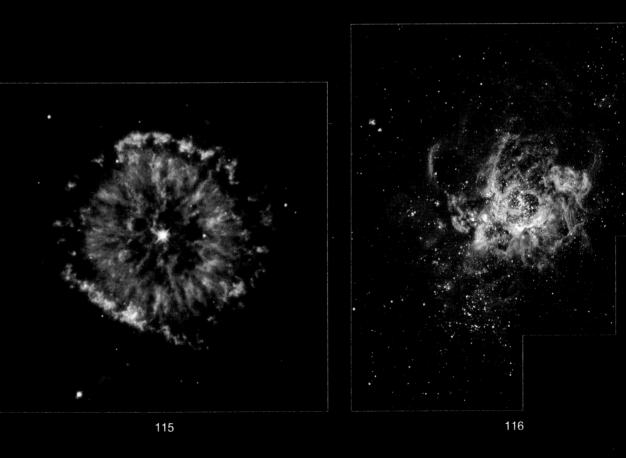

115

116

114. Carina Keyhole Nebula. **115.** Glowing Eye Nebula in the Constellation Aquila. **116.** Nebula in the constellation Triangulum.

117. Crab Nebula. **118.** Giant galactic Nebula. **119.** Pillars of Creation from Eagle Nebula.

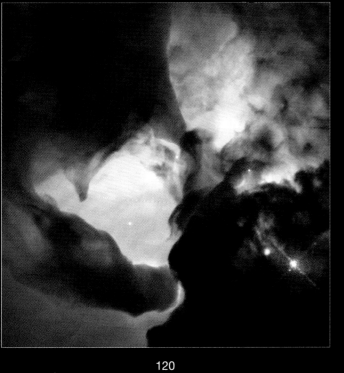

120

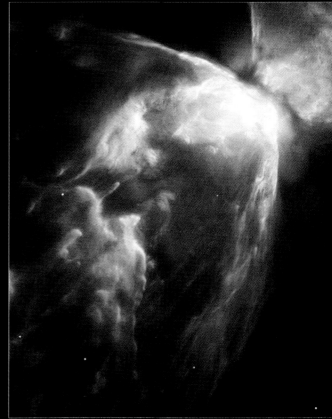

121

122

120. Giant twisters in the Lagoon Nebula. **121.** Bug Nebula, the brightest known planetary nebula. **122.** Eagle Nebula.

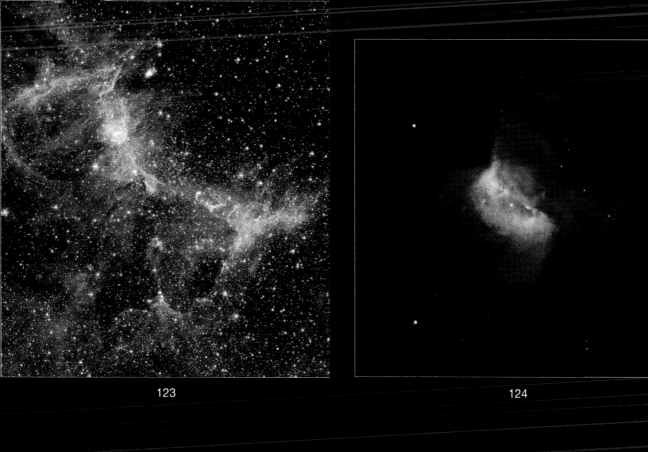
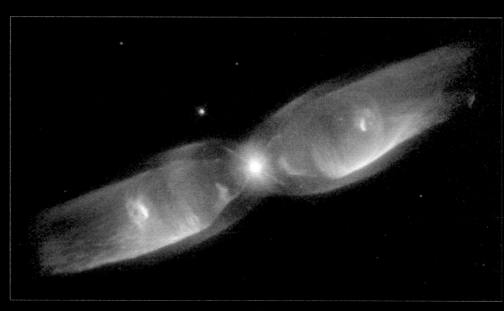

123. Eagle Nebula. **124.** Butterfly Nebula. **125.** Twin Jet Nebula.

126. Boomerang Nebula. 127. Cat's Eye Nebula in Constellation Draco. 128. Ancient stars in the Milky Way.

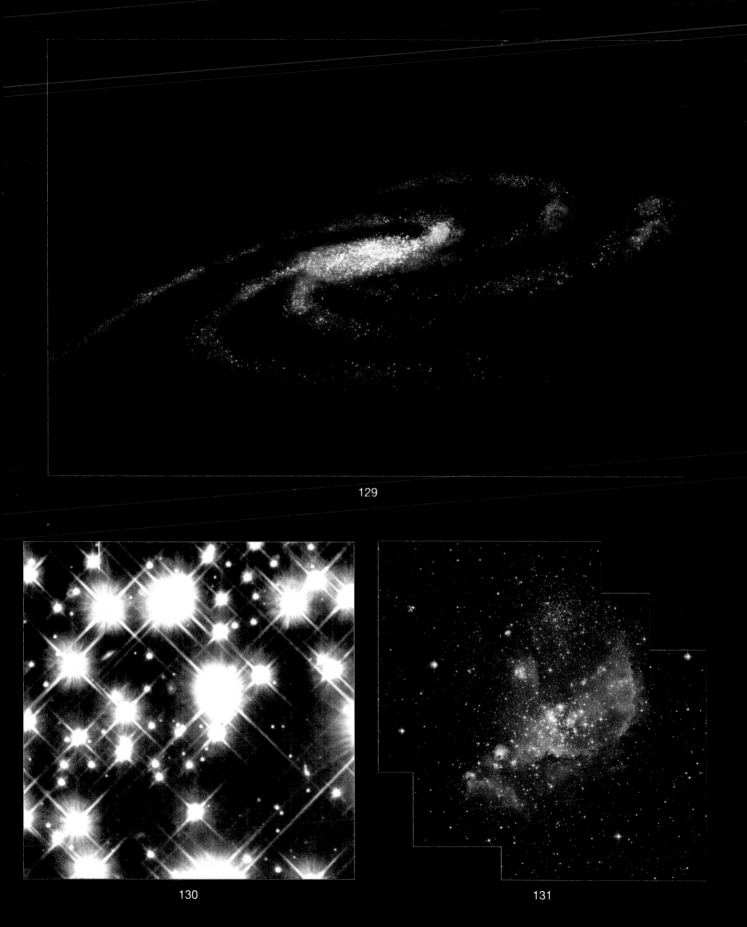

129. Milky Way galaxy. **130.** White dwarf stars in the Milky Way galaxy.
131. Embryonic stars in the Small Magellanic Cloud.

132

133

134

132. Whirlpool galaxy. **133.** Backward spiral galaxy. **134.** Spiral galaxy.

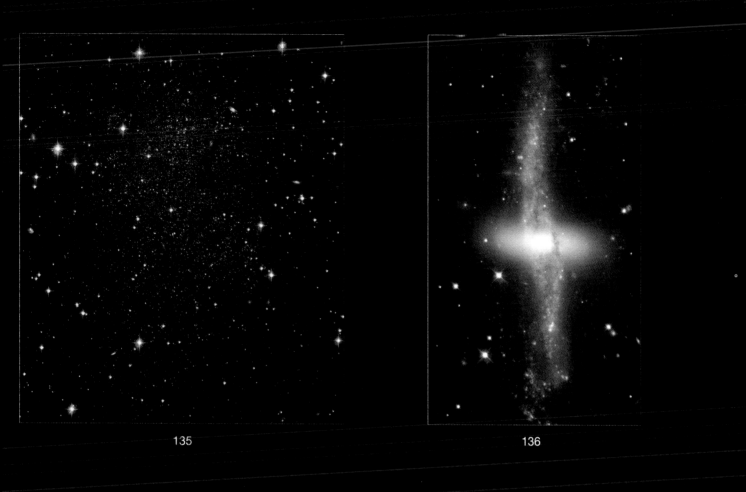

135. Sagittarius dwarf irregular galaxy. **136.** Ring around a galaxy. **137.** Ultraviolet light source in an old galaxy.

138

139

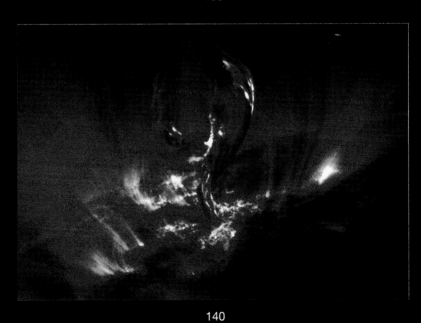

140

138. Sombrero galaxy. **139.** Andromeda galaxy. **140.** Coronal Mass Ejection.

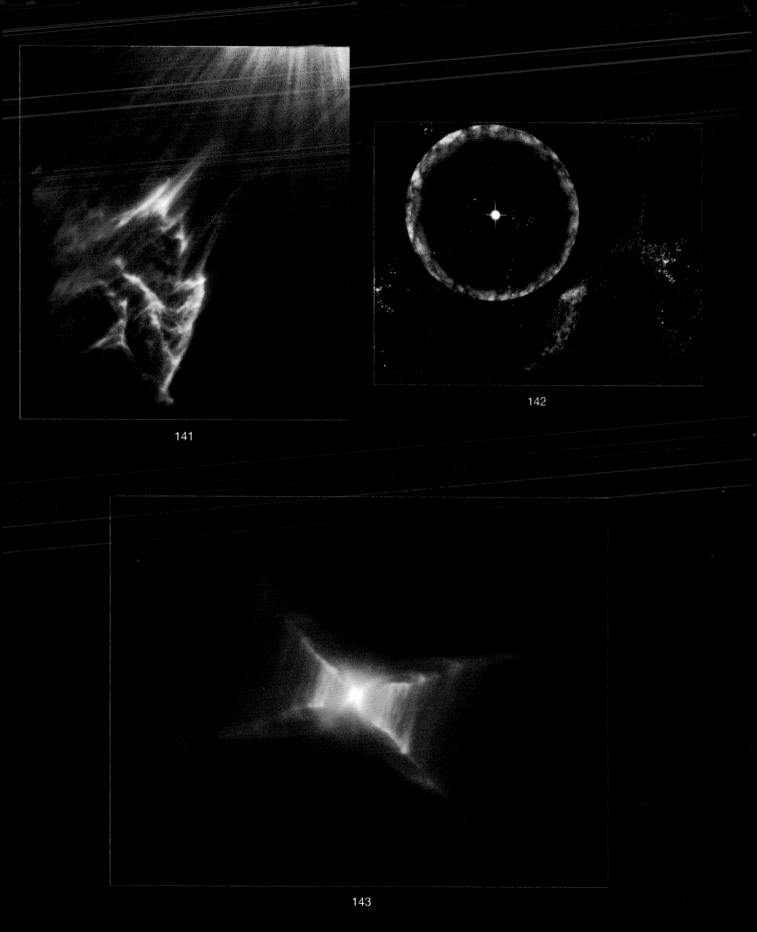

141. Barnard's Merope in the Pleiades Star Cluster. **142.** Stellar quake. **143.** Dying star, "Red Rectangle."

NEBULAS, GALAXIES, STARS, ETC. 45

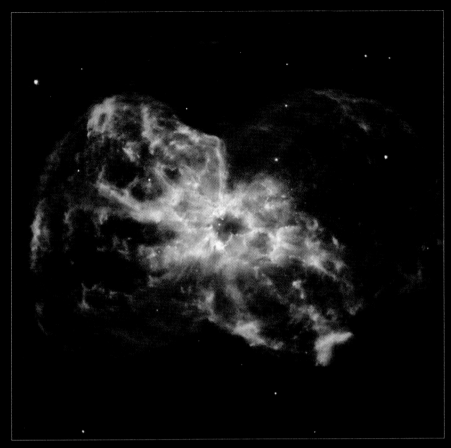

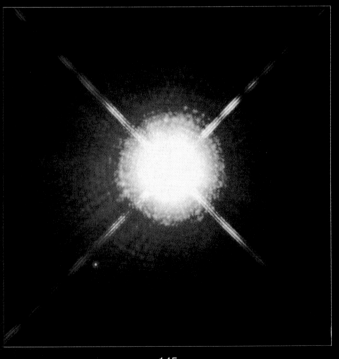

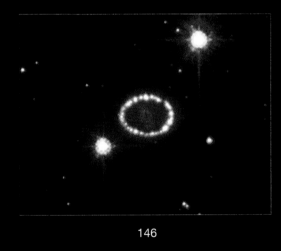

144. Dying star. **145.** Sirius, the Dog Star. **146.** Supernova.

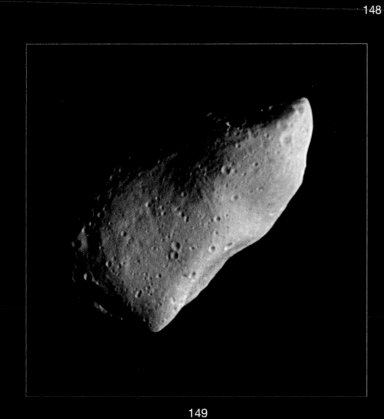

147. Icy comet NEAT. **148.** Celestial equivalent of a geode. **149.** Asteroid Gaspra.

150

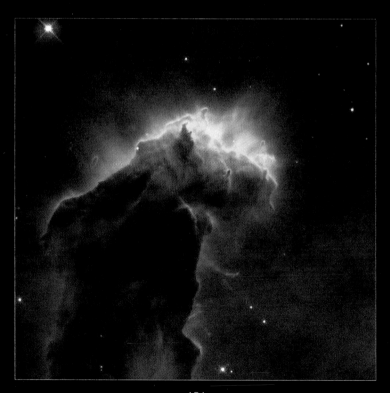

151

150. Cygnus Loop supernova blast wave. **151.** Molecular hydrogen gas and dust.